DMR FOR BEGINNERS USING THE TYTERA MD-380

BRIAN SCHELL

DMR FOR BEGINNERS
Using the Tytera MD-380

Copyright 2018 by Brian Schell.
All rights reserved, including the right to reproduce this book or any portion of it in any form.

Written and designed by: Brian Schell
brian@brianschell.com

Version Date: March 31, 2020.
ISBN: 978-1545553732
Printed in the United States of America

CONTENTS

Addendum/Update	1
Foreword	3
What is DMR?	5
Essentials of Using DMR	11
Programming the Tytera MD-380	13
Going Forward with DMR	33
Appendix A: DMR Self-Registration	37
Conclusion	39
About the Author	41
Stay Up To Date!	43
Help Me!	45
Also by Brian Schell	47

ADDENDUM/UPDATE

[Added March 31, 2020]

I have had emails pointing out that some of the images in this book are too small to read in some versions of the eBook (such as on the iPhone). This makes sense, since the book contains many screenshots taken from a full-sized PC screen.

In response, I have posted the full-resolution images from this book online at:

https://github.com/brianschell/DMR-Images

Good luck!

FOREWORD

In my other amateur radio books (D-Star for Beginners and Echolink for Beginners), I have tried to cover those topics as generically as possible, and most of what is said in those books applies to most or all of the radios that support those modes. This book is going to be a little different in that it *only* covers the Tytera MD-380 radio.

There are many varieties of radios, both for amateur use and commercial use, that support DMR. One of the most popular is the MD-380 by Tytera. This is mostly due to the price, as it's a good solid radio that can be gotten with most of the usual accessories (charger, programming cable, battery, belt clip) for under $120. There are vastly more expensive options out there, but from my experience there are more MD-380/390s in amateur/DMR use than any other radio. They're super common and quite affordable.

The other reason behind this choice is that there are vastly different ways of programming all the various radios, and there's no simple way to cover them all. I chose one common radio to cover, and I leave it to you to adapt or jump off into other radios if that is something that appeals to you.

And remember, the whole point of the book is to get you up and running quickly by eliminating the technical jargon and information you don't need to begin. There's a seemingly endless amount of technology and number of protocols and standards involved with the hardware and repeater side of DMR, and we're going to ignore a lot of that here. The book isn't about setting up your own repeater; it's about setting up your personal handheld radio and taking part in global communications. Once you're connecting and talking on the repeaters and with other DMR users, you'll find that there's still a lot more you can research and learn about that isn't included here.

Enjoy learning it all, and I'll talk to you on DMR!

>
> Brian Schell, KD8OTD
> KD8OTD@gmail.com

WHAT IS DMR?

DMR stands for **Digital Mobile Radio**, and is yet another digital radio standard, designed by the European Telecommunications Standards Institute. It's growing in popularity every year, and is one of the fastest-growing segments of handheld radio use.

Some Key features of the DMR standard are:

- Interoperability between manufacturers. Unlike D-Star, which was essentially locked into Icom for years, and Fusion, which seems to be the special pet of Yaesu, DMR radios are built by several competing manufacturers, from Motorola to Tytera.
- Backwards spectrum compatibility
- State of the art Forward Error Correction
- Analogue to digital migration with DMR systems
- Longer battery life
- 2 talk paths in a 12.5 kHz channel
- Doubling of capacity in existing channels

- Data applications interface (AIS)
- Two-slot Time Division (TDMA) operation

That's a mouthful. Basically, it's a way of merging amateur radio and digital communications through advanced repeaters that allow more than one channel of operation at a time. It is a competitor to D-Star, another digital voice mode that is also very popular.

The first major requirement is that in order to use any DMR radio, you *must* be a licensed amateur radio operator (ham). If you aren't a ham radio operator, then that's the necessary first step. In the USA, there are three main "levels" of license: Technician, General, and Amateur Extra. Technician is the beginning level, and it's not that hard to attain. The Technician license is all you need to do everything DMR offers. The higher-level licenses will allow you to do other fun things with your radio, but they aren't necessary to get all the benefits of DMR.

But I'm Not a Ham!

If you aren't already a licensed amateur radio operator, you will need to become one to proceed. A great place to start if you aren't already licensed is the **ARRL Ham Radio License Manual**. It's the book that gets most hams started. With that out of the way, everything else in this book will assume you are a licensed amateur. Again, it doesn't really matter which level of license you have for DMR, just get licensed.

What Can I do with DMR?

. . .

There are numerous benefits to DMR. First and foremost, it's all digital. With a good connection, this means you have crystal-clear audio and sharp voices all around. Or on a less-than-optimal connection, you get nothing but noise, often called "R2-D2" after the incomprehensible sounds of the famous droid from Star Wars. There are no half-heard or weak signals with DMR; it's all or nothing. It's not a perfect system, and it has its detractors, but then there are still hams out there who claim anything other than Morse code isn't "real radio." Ham radio is a big hobby, with lots of avenues to explore, and DMR is only one of them.

One of the primary disadvantages of DMR is that it wasn't designed with amateur radio operators in mind. DMR was created as a commercial solution for private companies that needed to talk long distance through their "walkie-talkies." DMR is a commercial system, designed and sold for commercial radios, and this fact impacts every aspect of DMR. If you wonder why it's so complicated to program these radios, this is the reason: it was meant to set up for a single channel or just a small handful of channels, and not to be changed much after that. Most DMR radios are sold for *either* UHF or VHF operation, while almost every other radio that amateurs use come in dual-band varieties. Again, it's because commercial radios don't need both bands, and it's a bit cheaper and simpler to make a radio that focuses on one band or the other.

A different digital radio system that was designed for HAM use would be D-Star, and the price difference for starter radios reflects that: D-Star radios start at around $350, while the Tytera radio that this book was written for is often found for around $120. That's not to say D-Star is *better* than DMR, but being a bit less complex and having fewer options means a lower price tag.

. . .

First Step: Registration

The absolute first step, regardless of how you plan to connect to DMR, is to get yourself registered with a subscriber ID. See Appendix A on how to do that. It may take a few days to get approved, so it's probably a good idea to do this right after you've ordered your radio. That way, your registration will be ready to go when you get it. It doesn't matter which DMR radio you plan to use, you need to be registered in advance for any of them, so do not skip this step.

Terminology

Throughout the book, I will talk a lot about repeaters and talk groups. It's easy to get the two mixed up, but they are not the same thing:

Repeater: A repeater is a device that sits high up, usually on the top of a tall building or tower somewhere. It accepts low-powered transmissions from nearby radios and re-transmits that signal through a more powerful radio signal and/or to somewhere else through the Internet. Hams use repeaters all the time on the VHF and UHF bands.

Talk Groups: Talk groups are unique to DMR. They are a lot like a party line or "chat room" where you can just talk to other hams without any concern for where they are. If you are familiar with other digital radio systems, a talk group is somewhat similar to an Echolink Conference or D-Star

Reflector. Talk groups have names such as WorldWide, WorldWide English, North America, Midwest, and other regional names. There are also "tactical" talk groups with names like TAC 310.

ESSENTIALS OF USING DMR

Getting Started

Before we get started actually programming your radio, we need to see how DMR handles commands and does things. This chapter explains the theory and process as generically as possible so you know what needs to be done and what information you will need to research on your own. The following several chapters are more hands-on in nature.

First of all, the MD-380 radio will work as a regular UHF or VHF radio, depending on which model you purchase. You can talk on the analog airwaves using either simplex communication or through a repeater just as you can with most other handheld radios. Although we will explain how to program in analog stations in a later chapter, the primary use of the radio is for DMR communications, so that's also the main focus of the book.

In order to talk on DMR, both the radio and the repeater you will connect to must support DMR. Although there are more and more repeaters that offer DMR, most of the repeaters Hams use are still analog, so before you even think about continuing, make sure you have access to a DMR repeater in your area. If you are close to a larger city, you probably have that access, but your mileage may vary.

Once you have your radio programmed to work with your repeater, you can make a call on a "channel." A channel is a combination of a frequency, a talk group number, a color-code, and a time slot. One of the unique features of DMR is that a repeater can allow two users to work the repeater's single frequency at the same time by using two "time-slots." If I am using time-slot 1 and you are using time-slot 2, we can both talk simultaneously through the repeater, as it "oscillates" between the two time-slots at a very high speed. Certain talk groups require the use of specific time slots, and different repeaters also use different "color codes" to differentiate between talk groups. It sounds like a lot, but we'll get into specifics in the next chapter.

Anyway, once you hit the PTT button and talk into the radio, your voice is encoded using all that information and sent to the repeater, which routes it through the Internet to all the other repeaters that are set up to use that talk group. If you key up your radio, you could be heard on literally hundreds of repeaters across the world. Someone out there hears you and replies, using their radio, programmed for their local repeater, and their voice is encoded and sent back to you.

Moreover, every radio/user is assigned a specific ID number. You can enter my ID number into your radio and call me directly or even send what amounts to a text message directly to my radio using my number. It's pretty neat!

PROGRAMMING THE TYTERA MD-380

The Impossible Way: Keypad

The Tytera (and all other DMR, radios that I've seen) require programming through a computer. There is no sane way to program a DMR radio through the number pad like you can with some other systems; there's just too many little details.

The Easy Way: Code Plug

Even considering the requirement of having a PC to program it, there is still an "easy way" and a "hard way" to program them.

The "Easy Way" to get your radio up and running is to install something called a "code plug." A code plug is a pre-made file containing all the settings and features for your local DMR repeater. Find your local Ham Radio Club's website or ask a member if they have a code plug for their repeater that works with your specific radio model. Some will have one; most will not. There are many different types of radios and situations. It's hard to find one setup that will please everyone, and radio enthusiasts are notoriously hard to please.

Still, if a code plug exists that matches your radio and repeater, it's a simple task to download it to your computer, load it into the software, and use the software to write it to your radio. If you have a code plug, you can skip through a lot of what follows, but it's still a good idea to get an understanding of how the radio works. If you don't have a code plug, or your code plug doesn't work exactly the way you want, then proceed to the next section.

The Hard Way: Software Programming

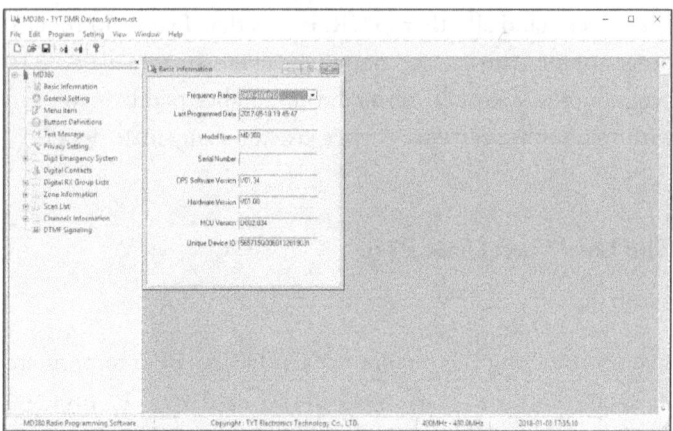

Figure 4.1: Basic Information

Sometimes you just want to manually enter all the data yourself. This requires a bit of research, planning, and data entry. It's not really all that hard, and it allows for complete

DMR FOR BEGINNERS USING THE TYTERA MD-380

control over what's in you radio--- you won't include any "useless" stations or programming that doesn't do anything in your local area. You can put in exactly what you need and only what you need.

The software required should be included with your radio, but if it's missing (or you bought a used radio), you can download it here:

http://www.tyt888.com/?mod=download

Once you have the latest version of the programming software installed and started, you'll see something like what you see in the image above.

There is nothing to fill in on the "Basic Information" screen,

Click on "General Setting" in the left-hand navigation pane and enter your Radio ID number that DMR-MARC sent you. See Appendix A if this doesn't sound familiar:

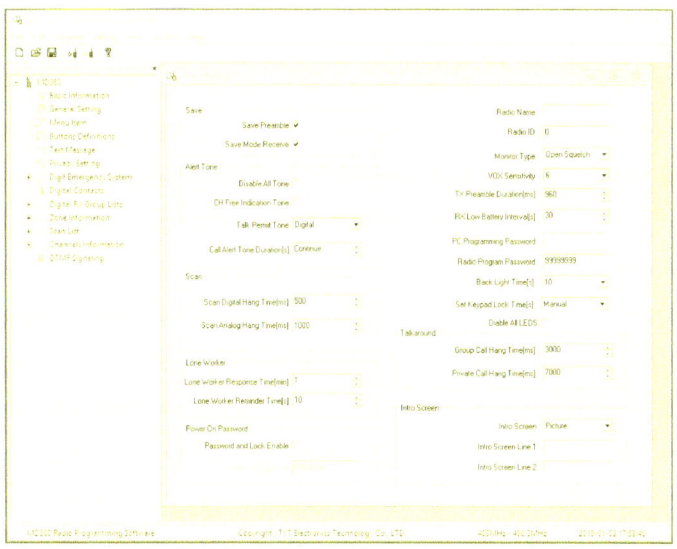

Figure 4.2: General Settings

OK, now we move on to actually programming the radio. It's a somewhat back-and-forth process that involves the following steps:

1. Get the programming information for your local repeater
2. Create a contact list of talk groups, people, simplex channels, and analog repeaters.
3. Create group lists
4. Fill in Channel Information for each talk group, individual, or simplex channel that you want to be able to use
5. Create zones for places you go
6. Create scan lists depending on how you want to scan using the radio.

Look up your repeater

Looking at the DMR-MARC web page for local Repeater Info, you'll have something that looks similar to:

DMR FOR BEGINNERS USING THE TYTERA MD-380

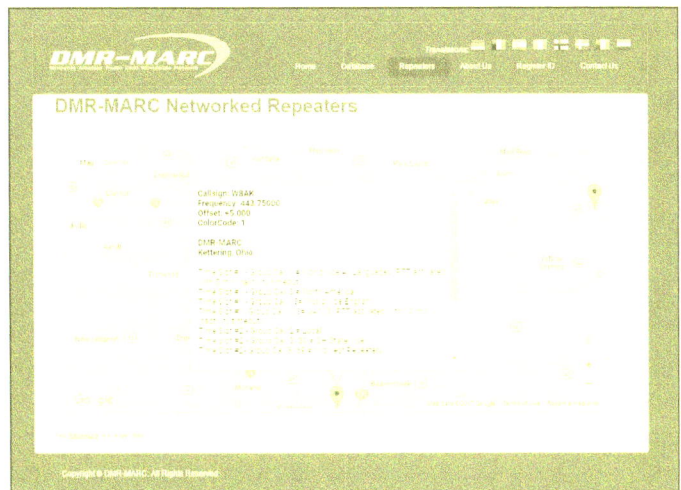

Figure 4.3: DMA-MARC Website

Make a note of all this information. For my local repeater, I see:

```
Callsign: W8AK
Frequency: 443.75000
Offset: +5.000
ColorCode: 1

DMR-MARC
Kettering, Ohio

Time Slot #1 - Group Call 1 = Worldwide All
Languages (PTT activated with 5 min inac-
tivity timeout)
Time Slot #1 - Group Call 3 = North America
Time Slot #1 - Group Call 13 = Worldwide
English
Time Slot #1 - Group Call 113 = UA113 (PTT
activated with 10 min inactivity timeout)
```

```
Time Slot #2 - Group Call 2 = Local
Time Slot #2 - Group Call 3139 = OH
Statewide
Time Slot #2 - Group Call 3169 = Midwest
Repeaters
```

In the MD-380 programming software, right-click on "Digital Contacts" in the navigation section on the left-hand side of the screen. Click the "Add" button at the bottom of the Digital Contacts screen. Use the information from the repeater's website to fill in the Contact Name and Call ID for everything the repeater offers. If you know there is a talk group that your repeater has that you aren't interested in, you can skip it. Most offer talk groups for WorldWide, Local, and other large regions.

If there is an individual PERSON, not a DMR talk group, that you wish to have identified when they make a call, you can add their name and DMR USER ID number as the Call ID here.

Also, it's probably a good time add a contact for making simplex calls. We'll talk about frequencies later, but for now, add a line called "Simplex" with a Call ID of 9 (in the USA and Canada). When that is done, close the "Digital Contact" windows. This isn't included on the repeater information screen because simplex channels aren't done through the repeater.

Here is the screen I filled out for my local contacts in Dayton, including my own call as an example of putting in an individual; you wouldn't normally make a contact for yourself:

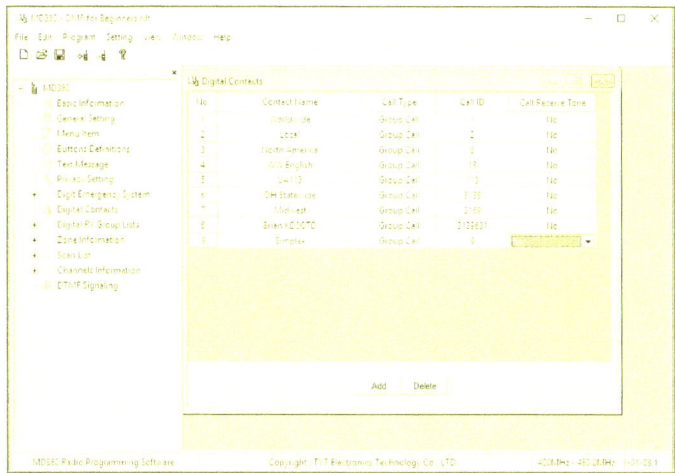

Figure 4.4: Digital Contact List

Next, click on "Digital RX Group List" on the left and click on "GroupList1." Rename it to be the same as the first contact name that you just entered in the "Digital Contacts." Click on the same name in the "Available Contact" side of this dialog and click the arrow to move it to the left. You have now created a RX Group for WorldWide:

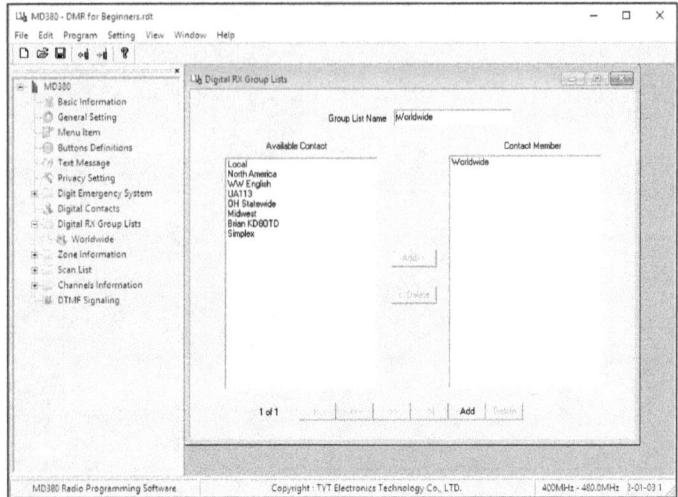

Figure 4.5: Group Lists

Repeat this process to create an RX Group List for each item in the Digital Contacts:

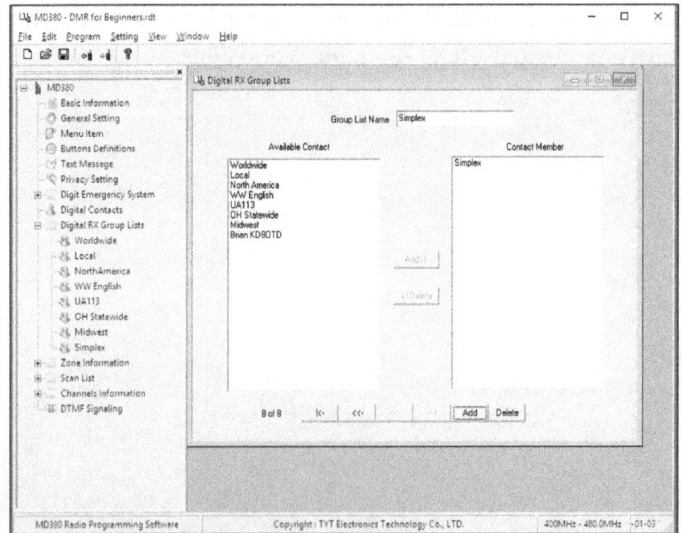

Figure 4.6: Many Group Lists (See left pane)

Once that is done, close the group list dialog and click on "Channels Information" on the left side of the screen. The blank channel information screen looks like this:

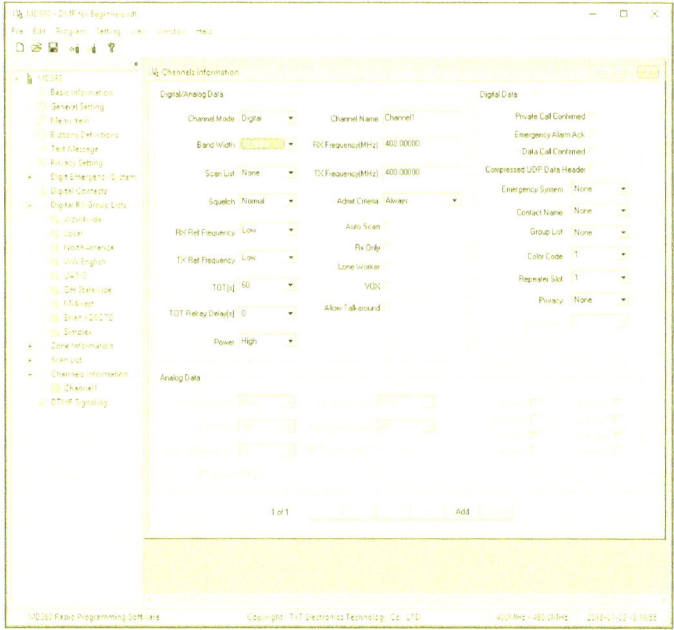

Figure 4.7: Channel Information Screen

Fill in the Channel Name for one of the group calls on the repeater information list. The first line of details on the W8AK repeater's information says:

Time Slot #1 - Group Call 1 = Worldwide All Languages (PTT activated with 5 min inactivity timeout)

I want to be able to use this talk group, so I fill out the "Channel Name" with whatever I like. I call it "W8AK

Worldwide". I then enter the receive and transmit frequencies in the appropriate boxes. It's also important to enter the correct "Repeater Slot" (1 in my case) and "Color Code" (also 1 in my case) for this repeater.

On the same screen, pull down the drop-down list for "Contact Name" and see all the digital contacts you entered a while ago. Since I am doing the talk group "Worldwide," I choose "Worldwide" from the list.

On the same screen, also pull down the drop-down list for "Group List" and choose whatever you typed for your matching group list.

Now the screen looks like this:

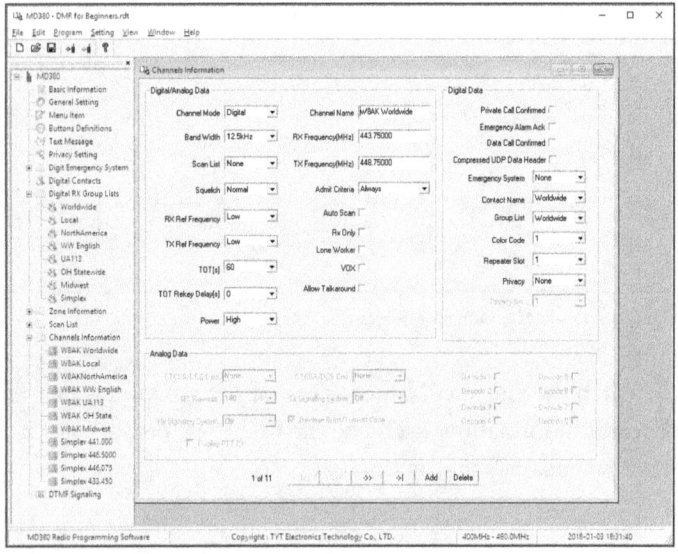

Figure 4.8: Channel Information for W8AK WorldWide

Don't change any of the other settings unless you know some reason why that is necessary.

Next, we will duplicate and edit this channel until we have created all the channels we want.

1. Over on the left-hand navigation pane, right-click on "Channels Information" and choose "Add" from the list of commands. A new channel will appear under the one you just made.
2. Next, right-click on the finished channel you just filled in all the information for and choose "Copy" from the list of commands.
3. Next, right-click just on the new, blank channel and choose "Paste." You have now copied the information from one channel to another.
4. If something above failed, try again, or just fill in the data manually for each channel that you want.

Fill in a separate channel for *each* talk group or option that your repeater offers. As shown above, my local repeater W8AK, offers Worldwide All Languages, North America, Worldwide English, UA113, Local, Ohio Statewide, and Midwest Repeaters. That's seven different things, so I created seven channels, one for each option. Be sure to use the correct "timeslot" for each channel. Some options are only available for slot 1, others require slot 2. The color codes for a repeater don't change.

Simplex Channels

While we are at it, let's add a few DMR simplex channels. Within the USA and Canada, the simplex frequencies are:

FREQUENCY	BAND	TALKGROUP ID	TIMESLOT	COLOUR CODE
441.0000	UHF	99	1	1
446.5000	UHF	99	1	1
446.0750	UHF	99	1	1
433.4500	UHF	99	1	1
145.7900	VHF	99	1	1
145.5100	VHF	99	1	1

Simplex frequencies for DMR

If you aren't in the USA or Canada, you'll need to look these up for your location.

Under "Channels Information" add blank channels and fill in the TX and RX frequencies for the ones that are appropriate for your radio.

As you can see below, I have seven talk groups and four simplex channels under the "Channels Information"

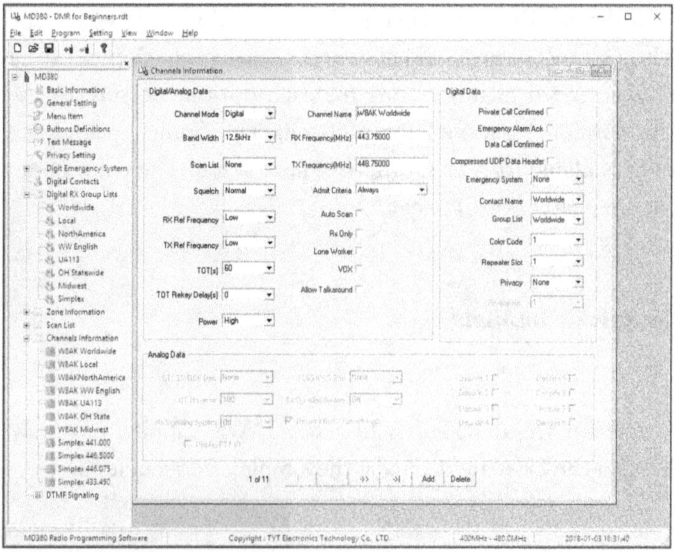

Figure 4.9: Added Simplex Channels

Zone List

Next, we will create a Zone List. In the navigation pane, click on "Zone Information" and click on "Zone1". You can rename this to anything you like. I put "Home" here, since this is the setup I would use from my home location. Use the dialog to add all your created channels to this zone:

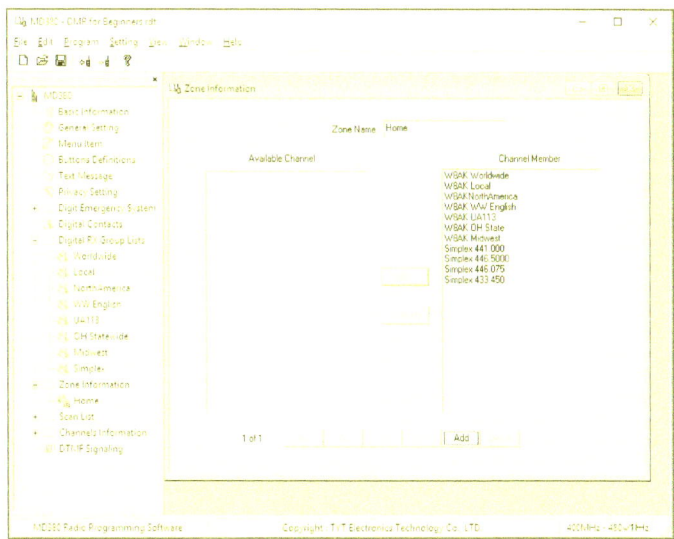

Figure 4.10: Zone List

If you travel a lot or have easy access to more than one DMR repeater, you can create a separate zone for each repeater and duplicate everything we have done so far for each repeater. You can easily choose which zone you want to use from the numeric keypad on the radio later. You should set up all the zones you might need on the computer and then switch on-the-go as necessary.

. . .

Scan List

Lastly, we can create a "Scan List." This one is optional, but nice. Click on "Scan List"in the navigation pane, then choose "ScanList1" from the list. You can rename this if you choose, and add whatever channels you wish to be able to scan to the list on the right. You can also choose priority channels here.

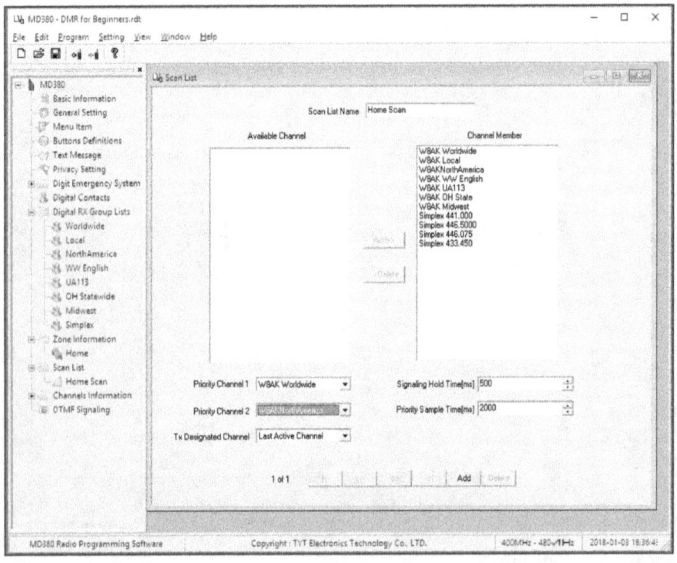

Figure 4.11: Scan List

Next, we need to go backwards a bit. Click on the "Channels Information" in the navigation pane to bring up the list of channels that you made earlier. Edit the setting for "Scan List" for each of the channels that you wish to be able to scan. You must do each one individually.

After everything is complete, you should have a selection of

- Digital Contacts
- Digital RX Group Lists
- Zone Information
- Scan List
- Channels

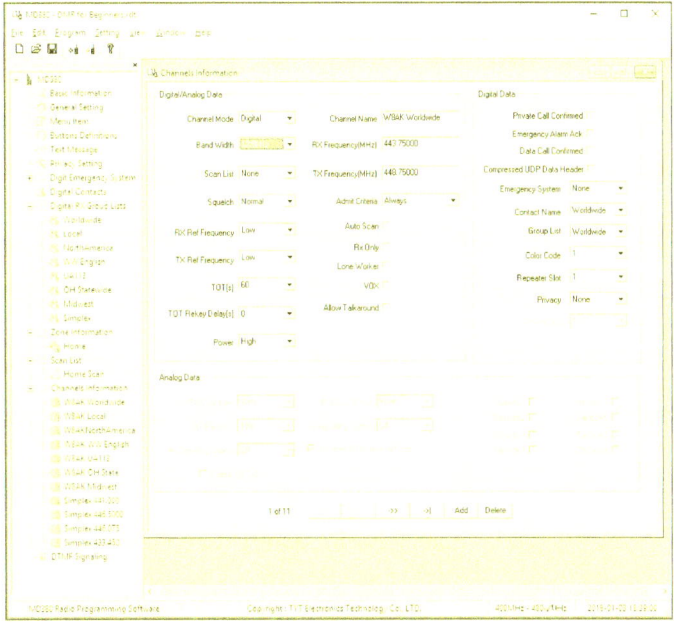

Figure 4.12: The Works!

Programming in Non-DMR Repeaters

While DMR is great, and probably the point of having this particular radio, it's nice to be able to talk to the regular "old-style" analog repeaters as well.

You'll need to find a list of your local area repeaters-- Your local ham radio club probably lists them on their website. You will need, at the very minimum, the receive and transmit frequencies and the PL tone, if there is one.

In the MD-380 software, add a new channel under the "Channels Information" section, just as you did with the DMR channels. Pull down the "Channel Mode" list and choose "Analog." All the digital information on the right section of the screen should immediately become grayed-out.

Enter the repeater name, RX frequency and TX frequency in the top section. Also, change the "Admit Criteria" there as well. In the bottom section, "Analog Data," enter the PL or CTCSS/DCS frequency if your repeater uses one. My local repeater is WB8VSU, at 442.300MHz, with a +5 offset (making the TX frequency 447.300) and a PL tone of 123.0. Here is how I entered all that:

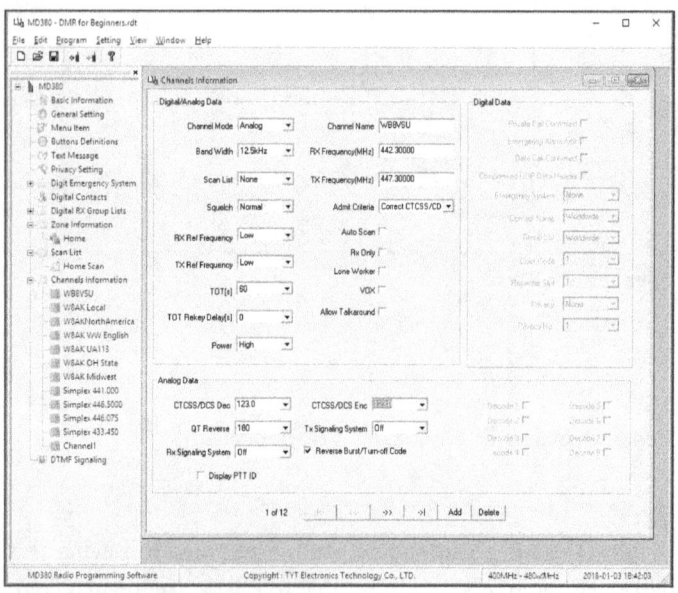

Figure 4.13: Added an Analog Repeater

Don't forget to add this channel to a zone and optionally a scan list.

More about Zones

Programming "Zones" into your radio allows you to have preset groups of channels for different repeaters or areas. I live in the Dayton Ohio area, so I have a zone set-up for that repeater, but once in a while, I also travel to Cincinnati and Lansing Michigan areas, so I have zones set up for the repeaters in those areas as well. By separating them into zones, I don't see those "distant" channels in my day-to-day usage. You set up the zones in the programming software; it's pretty straightforward after getting your first zone programmed in as done earlier. To switch zones on your radio, press the green button on the keypad to bring up the main menu. Choose "Zones" from the list using the up and down arrow keys, and then choose which zone you want the same way. Press the green button again to select it. Then choose your channel using the knob as usual.

Scanning

You can set up one or more "Scan lists" that are just what they sound like: lists of channels that will be scanned. Creating and populating these lists is done with the programming software. To make your radio perform the scan, press the green button on the keypad to bring up the main menu. Choose "Scan" from the list using the up and down arrow keys. You can then choose to either view the list of channels or choose "Turn On" to commence scanning. The radio will scan through the programmed channels and stop when it encounters someone transmitting. There is a small icon that appears in the top line of the display to show that the radio is scanning, but otherwise the screen shows the last channel. Note also that a scan list is limited to 16 channels, but you can create more than one list.

Tip

If you find yourself changing zones or scanning often, you can program these functions into the top or bottom programmable buttons above and below the PTT button. This is done via the programming software in the "Buttons" section of the navigation pane.

If you click on "Buttons Definitions" in the navigation pane, you can program various features for the various user-defined buttons.

Writing to the Radio

And that is all for the basic programming. Pull down the File menu and save your settings to a file somewhere.

For the final step, we need to write all this to the radio. Plug in the cable connecting your computer to your radio, and turn on the radio. If you are lucky, it'll just work; if not you may need to update your USB drivers. On my system, running Windows 10, it just automatically searched for the drivers it needed and worked fine.

Once you are confident that your computer can "see" the

radio, it's time to write the settings. Pull down the "Program" menu and choose "Write Data." You should see a dialog box that says, "Write data to radio" with a box for "OK" and "Cancel." It's not immediately apparent, but you'll need to click "OK" to start the transfer. You can then watch the progress bar as it writes the data to the radio.

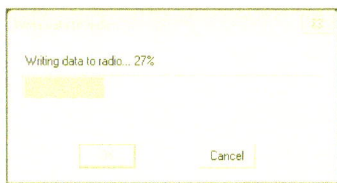

Figure 4.14: Writing to the Radio

Assuming all went well, you'll get a pop-up stating that "Write data succ!" No, that's not a typo-- If you ever needed proof that the radio was made in China, there it is. If you got an error, then there is probably a problem with the USB driver. With Windows 10, you can press "Windows-X" and click "Device Manager" to see your hardware devices quickly:

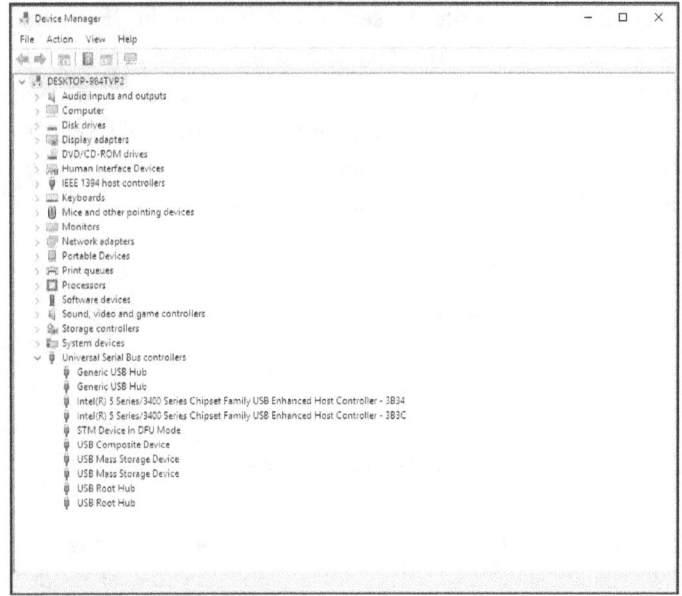

Figure 4.15: Windows 10 Device Manager

Your screen probably won't match that exactly, but if you see anything with an error message or highlighted in yellow, then that is likely where to problem lies. Right-click on the problem area and try to find updated drivers for your USB. Of course, make sure the cable is plugged in securely and all the other usual problem-solving steps.

Assuming it did work, you can unplug the cable from the radio. It's a good idea to turn the radio off and back on at this point, just to make sure everything resets properly.

GOING FORWARD WITH DMR

Here are a handful of additional resources and things to get involved with using DMR. These may or may not appeal to everyone, but if you want to do more with DMR than I've described so far, these various sites should be your next step.

DMR Nets

Nets are fun ways to listen to users around the world (or regionally, or even just locally) and talk about specific topics. These come and go, but here are a couple of web links to get you started:

http://trbo.org/nets.html
http://dmrworldwidenet.blogspot.ca/

Who's On?

See who else has been on DMR recently and where they are:

```
http://www.ve2tax.com/lastheard.aspx
https://ham-digital.org/dmr-lh.php
```

These links are a good way to see if there are any "busy spots" on the network. It's also a quick way to test your radio: key the microphone for a half a second (if no one's talking) and see if your callsign shows up on the list. If it doesn't, you might not be connected to the network properly.

Text Messaging

This is a fun one, and is unique to DMR. Back when programming the radio, I mentioned that you can enter an individual as a "Digital Contact." By pressing the green button on the keypad and choosing "Messages," you have a whole selection of options to choose from. You can experiment with most of them, but the easiest way to get started is to choose "Quick Msg," then use the keypad to enter a short message. Use the green button to "Send" the message. As a final step, you will enter the other person's radio ID number.

No Repeater? No Problem!

In some cases, you may have a local repeater that doesn't carry the talk groups or TAC channels that you want to use. You can't please everybody! Or perhaps even, worse, you don't have *any* DMR repeaters local to where you are. This isn't the unsolvable problem that it used to be. New products are coming to market that allow you to use very low power to connect to a "dongle-style" device that plugs into your home Internet connection instead of a local repeater.

Probably the best-known example of this kind of device for DMR is the SharkRF openSpot, pictured here. This device plugs into your Internet router and can be configured through your PC, tablet, or smartphone using a browser-based interface. This particular device can be used with DMR radios as well as D-Star and Fusion radios. Users can even use the Tytera DM-380 to connect to the Fusion network with this device. Previously, you would have needed to buy a Fusion-compatible radio to do this. For more information, check out my other book, **"OpenSpot for Beginners: D-Star, Fusion, and DMR Accessed Easily"** available wherever you picked up this guide.

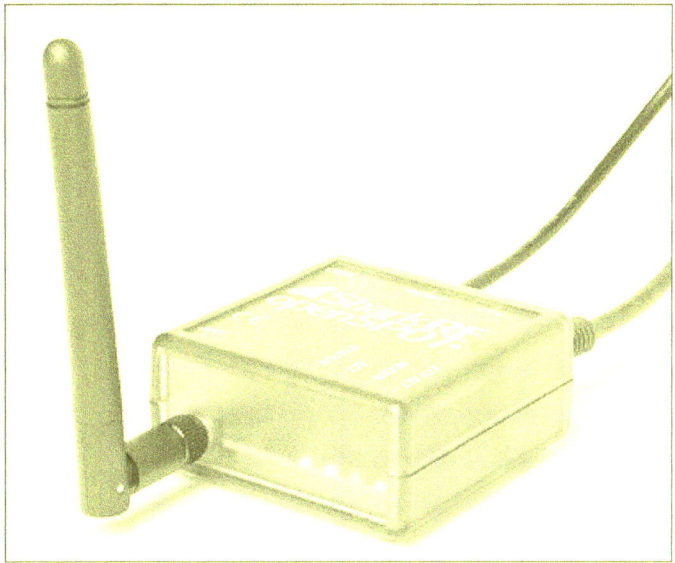
Figure 5.1: SharkRF OpenSpot

The primary disadvantage of using this kind of device is the cost. As of this writing, they run about $230, nearly twice the usual cost of the radio itself. Still, if you NEED the device or spend a lot of time using other digital modes in addition to DMR, it does exist and offers additional capabilities.

APPENDIX A: DMR SELF-REGISTRATION

The very, very first step in getting started with DMR is to get registered with DMR-MARC. You can do this even before you decide which radio you are going to buy. The organization that runs the site (The Motorola Amateur Radio Club DMR Group) is also responsible for assigning DMR ID numbers. This code number is essentially your "address" for using DMR, and you need one before you can do anything else. Registration is easy; just go to <http://dmr-marc.net/> and click on "Contact Us" and then "I'd like a USER ID for my radio." The next page will have some instructions and terms that you should look over. Once you have read through all this, click on "User Registration" at the bottom.

Figure 6.1: DMR User Registration

Enter your call sign in the box.

The site will look up your information and partially populate the next screen for you. You will need to fill in the missing information. The applications for the numbers are screened by an actual human being, so fill in the comments with something interesting about yourself.

Figure 6.2: Personal Information

Be careful entering the *Captcha* information in the graphic on the right. When you're done, hit submit, and usually within 24 hours, you'll receive an email containing your fresh new DMR ID number.

CONCLUSION

And that's basically it for getting started with DMR on the MD-380. We skipped over a few settings in the programming software that you may want to revisit and fiddle around with, but you are on the way!

If you come across any nifty tricks or things to do with your DMR radio, I'd love to hear about them, email me any time. If you have any suggestions for future additions to the book, there *will* be a second edition eventually, so let me know. Otherwise, catch me on the airwaves!

ABOUT THE AUTHOR

Brian Schell (KD8OTD) is a former College IT Instructor who has an extensive background in computers dating back to the 1980s. Currently, he writes on a wide array of topics from computers, to world religions, to ham radio, and even releases the occasional short horror tale.

He'd love to hear your stories of success and failure with the Tytera MD-380. If there's something you would like to see in a future edition of the book, or otherwise have suggestions, please drop him a note. Contact him at:

```
Web: http://BrianSchell.com
Email: brian@brianschell.com
```

- twitter.com/BrianSchell
- facebook.com/Brian.Schell
- amazon.com/-/e/B00B1E69J0
- instagram.com/brian_schell
- pinterest.com/brianschell

STAY UP TO DATE!

Join my email update list— There's NO weekly SPAM or filler material, only announcements of new books or major updates.

http://brianschell.com/list/

HELP ME!

CONTACT THE AUTHOR

If you have a suggestion or find a mistake, email me about it, and I'll get it into the next edition of the book. Got a gripe, complaint, question, or just adoring fan mail? Same thing!

LEAVE A REVIEW

If this book helped you, please leave a review where you purchased this book. Reviews are the best way to help out!

SHARE WITH YOUR FRIENDS

Did you enjoy this book? Please use the buttons below to spread the word to your friends and followers.

ALSO BY BRIAN SCHELL

Classic Horror Films:

- Horror Guys Guide To Universal Studios Shock Theater
- Horror Guys Guide To Universal Studios Son of Shock!
- Horror Guys Guide To Hammer Horror

Old-Time Radio Listener's Guides

- OTR Listener's Guide to Dark Fantasy
- OTR Listener's Guide to Box 13

Amateur Radio

- D-Star for Beginners
- Echolink for Beginners
- DMR for Beginners Using the Tytera MD-380
- SDR for Beginners with the SDRPlay
- Programming Amateur Radios with CHIRP
- FM Satellite Communications for Beginners
- Trunking Scanners for Beginners Using the Uniden TrunkTracker

Technology

- Going Chromebook: Living in the Cloud
- Going Chromebook: Mastering Google Docs
- Going Chromebook: Mastering Google Sheets
- Going Chromebook: Mastering Google Slides
- Computing with the Raspberry Pi: Command Line and GUI Linux (Technology in Action)
- Going Text: Mastering the Power of the Command Line

- Going iPad: Ditching the Desktop
- DOS Today: Running Vintage MS-DOS Games and Apps on a Modern Computer

The Five-Minute Buddhist Series

- The Five-Minute Buddhist
- The Five-Minute Buddhist Returns
- The Five-Minute Buddhist Meditates
- The Five-Minute Buddhist's Quick Start Guide to Buddhism
- Teaching and Learning in Japan: An English Teacher Abroad

Fiction with Kevin L. Knights:

- Tales to Make You Shiver
- Tales to Make You Shiver 2
- Random Acts of Cloning
- Jess and the Monsters

www.ingramcontent.com/pod-product-compliance
Lightning Source LLC
Chambersburg PA
CBHW061223180526
45170CB00003B/1137